NEW BEEKEEPING
in a
Long Deep Hive

Beginner beekeepers inspecting a nucleus at the rear of a Dartington Long Deep hive at HoneyWorks Beekeeping Training Centre, Hitchin, England.

NEW BEEKEEPING
in a
Long Deep Hive

by

Robin Dartington

With a foreword by
Cecil Tonsley, FRES

First published 1985
Revised edition 2018
© Robin Dartington 2018

ISBN: 978-1-912271-19-1

Published by:
Northern Bee Books, Scout Bottom Farm
Mytholmroyd, Hebden Bridge
HX7 5JS (UK)

www.northernbeebooks.co.uk

Tel: 01422 882751

Front cover shows Dartington Long Deep hives in the author's Radburn Apairy in Letchworth Garden City, Hertfordshire, England.

Rear cover shows mix of Deep Long hives and National hives at HoneyWorks Beekeeping Training Centre, Hitchin.

Full details for making Dartington hives at home are given in **Construction Information for Dartington Hives**, Robin Dartington, Northern Bee Books 2018.

Fuller information on new ways to enjoy keeping bees using Dartington hives will be covered in **The Manual of New Beekeeping**, to be published by Northern Bee Books in 2018.

Printed by Lightning Source, UK

FOREWARD
TO THE 1985 EDITION

I have great pleasure in writing a foreword to this short book in the hope that it will stimulate its readers to consider principles and to make their own experiments along the lines of the author. The methods of management are not original ways of managing bees, as the author admits, but are not conventional.

Long Idea Hives, Coffin Hives and Long Deep Hives are of course all known to bee keeping. In Hungary, for example, many bee keepers use a Long Deep Hive quite successfully, but do not manage them on quite the same lines as those put forward by Robin Dartington, whose main intention is to interest the bee keeper who keeps a few hives in his garden and who needs a relatively easy and reliable form of swarm control.

In this work, the author explains his own use of a particular pattern of long hives which he has made to his own design, but which is derived from standard lines. The hives have been in use for ten years, which some bee keepers will say is not long enough to prove its efficiency. However, many different systems have been advocated and explained to bee keepers over the years, and how many of those systems, however long in use, have proved 100% efficient? Bees, like many things in nature, are compelled to adapt to their behaviour greatly to suit the varying climate year by year.

This book has been published to make bee keepers think, and I believe it may do just that.

Cecil Tonsley

PREFACE
TO THE 1985 EDITION

This booklet describes a new approach to keeping bees, which has given me a lot of enjoyment, and I hope it will interest other people who keep bees either as a hobby or for profit. The order within the bee hive has fascinated mankind for hundreds of years and the search for better ways to manage bees always continues.

This system needs a long hive body which takes deep frames. The British Standard Extra-Deep 14" x 12" frame is ideal. There is no need to move the hive during the season; if 'artificial swarming' is necessary it is achieved by opening a rear entrance and not by placing the whole hive in a new position. So the overall weight is unimportant, and the hive can be positioned permanently on rails at a comfortable working height.

There are sound objections to a 'long-idea' hive if it is used with standard frames. When fewer extra-deep frames are used together with shallow supers, the system conforms with good principles. Following the introduction of a new type of extractor by Brinsea Products, the extra-deep frames can be extracted easily and without breakage.

I am grateful for help from John Kinross of Bee Books New & Old, and from Jeremy Burbidge of Northern Bee Books. Almost all the ideas here were found individually in old books, but they seem not to have been put together before in one system.

This booklet deals both with theory and practical management. How to make your own hives is described in Construction Information for Dartington Hives, published by Northern Bee Books. Anyone wishing to try the system is welcome to contact me.

Robin Dartington   20 July 1985

PREFACE
TO THE 2017 EDITION

The first edition in 1985 introduced the Long Deep hive to British beekeepers and led over the following 30 years to giving talks to associations between Cumbria, Kent and Cornwall. Some beekeepers were persuaded to give the hive a try – but as new beekeepers need to learn the craft through a training course – and as tutors naturally teach what they are familiar with – it has not become mainstream.

Beekeeping – and in particular beekeepers – have however changed over the last 30 years. Growing affluence means that fewer beekeepers look for a sideline income from bees - more are attracted to the fascination of exploring this ancient and complex life-form as a way to contact nature, a change from recreations using a phone, tablet, computer, or TV. The advantages of a Deep Long hive - safety, convenience, economy and style - are relevant to these new beekeepers – they are not so physically strong, they value neatness more, they do not want to move their hives around to get a bigger honey crop. So I am grateful to Northern Bee Books for re-issuing this little work to interest a new generation.

Beekeeping itself is changing with the growth of interest in more 'natural' ways to manage bees – more respectful of the ways bees have found to survive for millions of years. In nature, the honeybee nest is round – the brood nest fitted well into round skeps and the domed roof meant that the dwindling cluster was still tightly enclosed as it climbed up its diminishing winter stores. Cubic wooden hives are the product of technology, mechanical saws, not of nature. Kenyan Top-bar hives taper downwards to make a better fit with natural combs and the Sun Hive seeks to both encircle the nest and exactly clothe the natural catenary shape of the combs. Hives holding rectangular frames cannot do that – but a Long Deep hive can easily be fitted internally with just top-bars and sloping sides or even a catenary shaped liner – which is where my own experiments are heading.

The Long Deep Top-Bar hive made in 1980, with a side observation panel, showed me that bees respond well to this shape of cavity – and that the principles of management set out in this booklet in 1985 can still be applied.

This reprint is in facsimile as the computer died long ago and so some references cannot be updated, for which you must excuse me. For example, the roof of a long hive is now best made in two pieces – and the advertised Honey-Twin-Spin extractor is no longer available. A pity, as it took two shallow frames lying flat in a fibreglass bowl set on a table top – light, easy to use, clean and store - quite sufficient for the hobbyist with only two or three supers to empty. Perhaps the illustration here will lead to its re-introduction!

Enjoy your 'new beekeeping'!

Robin Dartington

CONTENTS

Chapter   Page

Foreword to 1985 Edition

Preface to 1985 Edition

Preface to 2018 Edition

Introduction 1

1. Origins 3

2. Observations 9

3. Reflections 15

4. Equipment 22

5. Operations 25

6. Conclusions 36

References 39

INTRODUCTION

New Beekeeping is a break-away from the conventional approach to beekeeping. The essence is attention to the individual, — the life behind the phenomena.

New Beekeeping examines the 'life urge' in the bee colony, (centred on the queen), rather than mechanical colony behaviour.

Conventional management aims to master the honey bee through physical domination (we disrupt and frustrate the desire to swarm). New Beekeeping seeks to understand the urges to which the bees are subject. We can direct the colony to by-pass the swarming urge without need to demoralise or weaken the colony.

The underlying theory is that the 'state-of-mind' of a bee colony is not invariable, but progresses through a sequence as the colony develops. The bees' behaviour at any time depends on the state the colony has reached. Manipulations can be used to induce change from one state to another to suit the beekeeper.

The first chapter outlines conventional beekeeping for comparative purposes and explains the origin of the Long Deep Hive. Chapter 2 observes the growth of a new colony and in Chapter 3 the lessons are turned to advantage for practical management.

In Chapter 4, I outline the minimal equipment required for this system, which will be more fully explained in a further booklet.

Chapter 5 concerns practical management for explanatory purposes. I have set out a "Twelve Trip Plan" which will explain how the basic method of management can be completed in a "normal season".

In Chapter 6, I attempt an assessement of these ideas - even though that is premature. I should wait in hope to hear from my readers and fellow experimenters!

Chapter 1

ORIGINS

101. Conventional British hives are all designed to take either 10 or 11 BS frames 14" x 8½" deep in the brood box. One box allows only enough breeding space (together with stores and pollen) for a queen which lays up to 1800 eggs per day. The queens of strains prevalent today are capable of perhaps 1800 to 2000 per day, but probably only exceptional queens exceed 2000 per day. So two standard brood boxes are **therefore** generally too large, and one standard plus one shallow is a common compromise.

However, **few** beekeepers would deny that any brood nest in two boxes is not easy to manage. The bees need to keep the **central** gap at breeding temperature, and fiercely resent the boxes being separated. In addition the space is repeatedly bridged with brace comb, since the bees **find** a gap within the brood nest unnatural.

102. If a new stock formed **from** a nucleus is given ample additional super space in good time, it is unlikely to swarm in its **first** season. The next year however, the beekeeper must be ready **for** swarm preparations. Removing the surplus honey the previous autumn is in **fact** the start of swarm control **for** the following season - although of course the beekeeper has quite a different motive when taking the harvest!

103. In towns and suburbs, the beekeeper cannot risk worrying neighbours by letting swarms escape. Various ingenious methods have been devised whereby the natural urge can be **satisfied** without the bees swarming out of the hive.

Amateur beekeeping still owes much to T.W. Cowan, Chairman of the British Beekeepers Association **from** 1874 to 1922. His guide bpok of 1881 remained in print for 55 years. The method of **artificial** swarming he most strongly recommended was to put 4 or 5 brood frames in a **small** ('nucleus') hive and raise a new queen; then to substitute the nucleus **for** the old queen. Cowan called this 'nucleus swarming'. The method is still widely used, but the extra equipment is untidy in the garden, and the system also tends to create spare nuclei which all too easily grow into unwanted extra colonies.

The need **for** special nucleus hives can be avoided by using additional **full** sized brood boxes, and if these are placed over a stock (on the same stand) it becomes easy to combine the new and old units later in the season and so avoid an overall increase.

Very many systems on these lines have been devised by different beekeepers, which are described in Wedmore (1932) and Bent (1946). The typical manipulations for a single walled hive are:

Fig i: The colony is wintered on brood boxes B1 and B2: supers S1 and S2 are given by mid May.

Fig ii: When the brood nest is fully expanded the queen (Q) is transferred on one frame to the additional brood box X filled with empty comb; brood boxes B1 and B2 are replaced over the supers. The flying bees collect in box X, and are thereby temporarily separated from the brood (thus imitating swarming).

The nurse bees in the brood, being at a distance from the queen, will raise queen cells under the supercedure impulse.

Fig iii: One week later the shallow brood box B2 is replaced below the supers; a spare cover board is used to isolate the remaining brood and so form a nucleus with its own top entrance. The weakened nucleus will allow only one cell queen to mature.

Fig iv: When B1 has a new queen (q), boxes B1 and X are interchanged and the old queen (Q) is removed. A week later any queen cells in X are destroyed, and the two parts of the colony are re-combined.

This system will build a large foraging force, and will prevent swarming for the rest of the season. There are however two drawbacks:

a) In Fig ii, the Queen must be found amongst 22 brood frames, (this can be difficult if the queen 'runs' when the colony is disturbed by splitting B1 from B2).

b) The operations in Fig ii, iii, and iv each involve lifting a full standard brood box (weight 50 to 60lbs) up and above the supers.

These difficulties, plus the cost of the additional brood box, undoubtably induce many beekeepers to leave their bees alone and to pretend that they rarely swarm!

Fig i

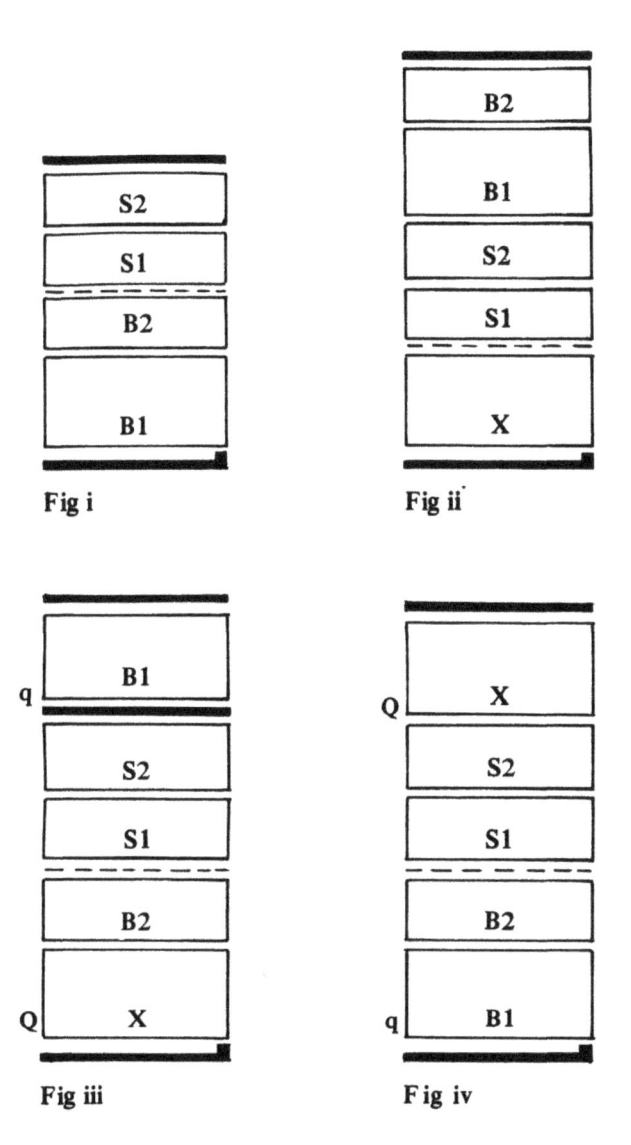

| S2 |
| S1 |
| B2 |
| B1 |

Fig ii

| B2 |
| B1 |
| S2 |
| S1 |
| X |

Fig iii

| q | B1 |
| S2 |
| S1 |
| B2 |
| Q | X |

Fig iv

| Q | X |
| S2 |
| S1 |
| B2 |
| q | B1 |

104. The first step to a better system is a change to BS 14" x 12" extra-deep brood frames, each single frame being equivalent to one standard and one shallow frame combined. The standard brood box can be simply deepened with a 3½" high 'eke'. It is quite impractical to 'elevate' the 80lbs weight of such boxes during manipulations, — but they can be dragged sideways.

The manipulations then become:

Fig v: Two supers (S1+S2) are added above the single brood box (B) by mid May.

Fig vi: When the nest is fully expanded, the queen (Q) on one frame is transferred to a new box and floor X, and box B is pulled away and turned about. All the flying bees return to box X, and box B (being queenless) will raise queen cells.

Box X is provided at first with only 6 empty combs (plus the queen on one brood frame); after 1 week it is strenghthened with 4 frames from box B, (box B becomes a six-frame nucleus).

Fig vii: When box B has a new queen, boxes B and X are interchanged, and frames are transferred once again so that box B is full again. After the flying bees have left box X, the old queen (Q) is found and removed. Queen cells in box X are removed after a week, and X is linked to B via rear openings in the floors. Box X will become deserted and it is then removed and stored for the following year.

This system is easier for the beekeeper. Only eleven frames need be searched for the queen, and the search has been made easy by the beekeeper waiting until all the flying bees have left the brood. There is however still the expense of a second deep brood box, cover and roof, which are used for only one month a year.

A variation which reduces the need for extra equipment is to use two shallow boxes as the temporary second brood box — these can be set over the hive at the end and used for honey storage (a total of four supers can be essential in some years).

Fig v

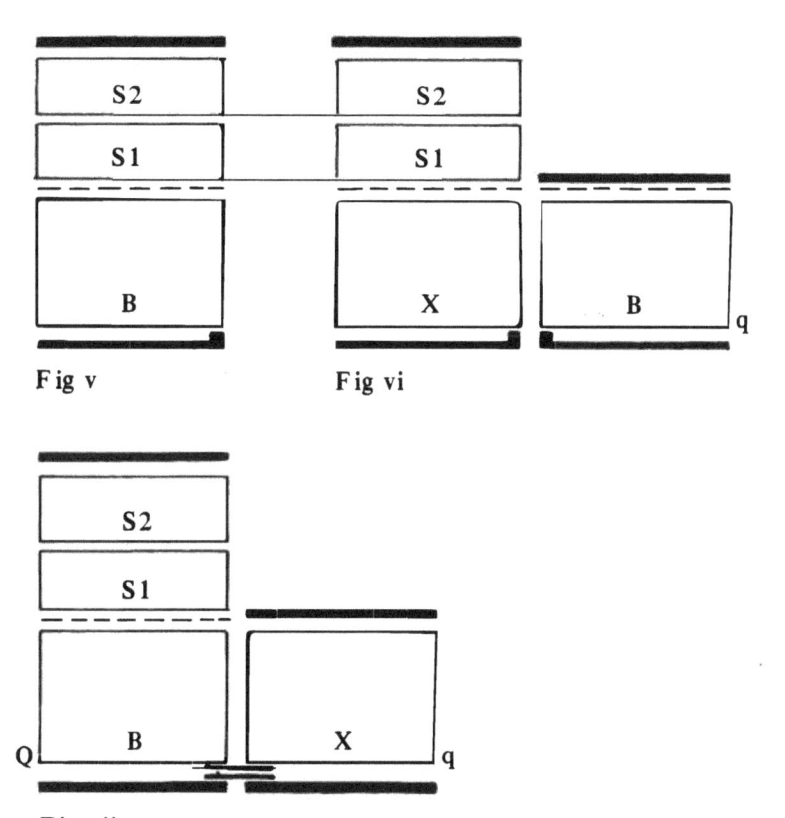

Fig vi

Fig vii

105. The system has however one serious weakness in that the new queen was raised from an 'emergency cell' within a nucleus. Queens should be raised under the supercedure impulse, and within the strongest of colonies.

My next step therefore was to combine the two deep brood boxes into one double-length, double-ended brood box — a 'long-deep' hive which could be adjusted in size or split into two separate chambers merely with a simple close-fitting 'divider'. Within this hive the brood could be spread laterally to induce the raising of a new queen under the supercedure impulse but it still formed one unit during cell raising. (See Fig viii).

This is the hive I have used over the last ten seasons, and it has proved a delight to operate. My methods have evolved through the years as I experimented. Last year, I gave away my last two 'vertical hives' when I realised I no longer had any wish to 'pull apart' a strong colony in a National hive.

Fig viii

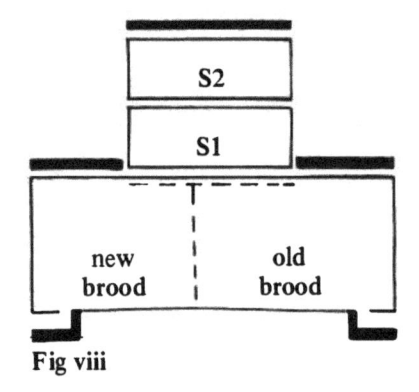

106. At first, I still operated the colony mechanically — ie. move A to B, then do C, — so altering the conformation of its nest without appreciating the natural response of the colony in its prevailing state.

However the manipulations were so much less disruptive to the nest that I found myself observing the bees more closely and reflecting on their behaviour throughout the season. I came to wonder if perhaps the proper way to approach manipulations was to seek to mould the bees' response to the changing seasons, rather than to frustrate their natural urges.

My interest increased when I found that the concept of a 'Hive Mind' had been discussed by Wadey (1946). New Beekeeping describes the theory and management system that evolved, which aims at manipulating the Hive Mind.

107. This brief account has explained that the Long Deep Hive originated through successive attempts to by-pass disadvantages in conventional systems. I have assumed that beekeepers agree on the following 'good principles' for any system for managing bees:

1. Colonies should be kept strong throughout the year. Only strong colonies can collect large harvests, and strong colonies are best at resisting disease.

2. Positive action should be taken to avoid the loss of swarms; from trial and error it appears that the best systems involve temporary separation of the flying bees (the swarm) from the brood (the parent stock) but ideally the temporary condition should not require the use of much extra equipment.

3. Manipulations should disturb bees to a minimum.

4. Queens should be replaced after their second season, but better still within their second season.

5. New queens should be raised by strong lots of bees, but should preferably be mated and tested in nuclei.

6. Brood combs should be renewed periodically (say after about seven years use).

Chapter 2

OBSERVATIONS

201. In Northern latitudes, wild honey bees generally live in the hollow trunks of trees, and so build vertically elongated nests to fit the cavity. This has perhaps influenced northern beekeepers to use vertically extending hives. The use of vertical cavities is not universal however, and around the Mediterranean and into Africa, native people generally keep bees in horizontal cavities.
Experiments by Seeley and Morse (1978), showed that bees in northern latitudes have in fact no preference for cavities which are vertical rather than cubic, provided the cavities are of the right size, dry and defensible.

202. In developing a new hive, I felt it might be worthwhile to observe how a colony left on its own would use a cavity that was deep but extended horizontally. I hived a swarm in a special observation chamber of the proportions of a Long Deep Hive but fitted only with top bars and starters. The new colony had to build its own combs. The arrangement was similar to the Kenyan hive described by Shida (1976), but was longer and not so wide.
The colony developed to full size over two years, building eleven combs each averaging 14 inches in width by 15 inches deep. It did not grow steadily, but in bursts of activity which seemed to be a response to the seasons outside. The first summer it showed no sign of swarming, but from late spring in the second year the colony had the appearance of being 'ripe'. I robbed it of its flying bees (by relocating it at midday), and then it appeared to settle back to rebuild its strength.
The colony built its combs parallel to the entrance, and at first these were right against the entrance wall. I moved the bees and combs behind a solid dummy, to leave a space about 6 inches (150mm) wide inside the entrance. It later built a twelfth comb at the rear of drone cells. Twelve seems the maximum it wants to build for normal use. However, while 'ripe', a cluster formed in the front cavity and built three combs perpendicular to the front dummy.
The first conclusion was that a Long Deep Hive is certainly not suited to 'leave alone' beekeeping. It is essential to spread the colony by spreading the frames once they are occupied by bees. Whereas bees may not build new combs more than fifteen frames from the entrance, experiments show that bees will not abandon occupied combs which are moved back within the hive.

203. I made almost daily casual observations, and reached the second conclusion that the colony had developed through a sequence of stages, or 'states-of-mind', which had controlled its behaviour.

Even after this one simple set of observations, I feel able to propose that there are nine 'states-of-mind' of practical importance.

State 1. The Loose Swarm.

A new colony starts as a loose swarm containing the parent queen and bees of all ages, but with a predominance of 'wax makers' (10 to 20 days old). The loose swarm has only one concern — to find a nesting cavity. Scouts communicate their different finds with dances, and the 'colony mind' selects one opportunity.

The mood of the 'loose swarm' persists until the queen has come back into lay. If meanwhile the scouts find a better cavity, the swarm may again take to the air.

State 2. Nest Founding.

Founding the new nest proceeds very rapidly. A single comb will be started first, and when this is perhaps 5 inches deep and wide, work starts on a second and third to each side.

As soon as cells are drawn out, the queen begins to lay. In this state, the colony is in a hurry. It will be 21 days before the first workers emerge, and meanwhile the population will dwindle steadily.

A strong swarm may build five natural combs perhaps 12 inches deep and wide before it turns its attention to survival. It has to store at least 25lbs of surplus honey for the coming winter.

Assuming a good nectar flow, the colony may later build a sixth comb. This will be done for storage and may therefore be of drone cells, which need less wax.

<u>Superceeding the Queen</u>

The stress of establishing a new nest will have taken its toll, and the colony may superceed an old queen at the end of its first season.

I do not know if a stock will superceed only while it has fertile drones. If so it may be August before this new colony is ready — and by that time other colonies may have rejected their drones with the consequence that the stock will breed true. Perhaps however foreign drones may seek out and join a stock in this state.

A colony undergoing supercedure may have no brood, and may seem to have no queen. The bees however do not exhibit any restlessness: and they would reject a foreign queen.

Supercedure is a process of regeneration which occurs whenever necessary. I doubt if it is itself a 'state-of-mind' of the colony, but once there is a new queen this will effect later behaviour.

State 3. Immature Wintering,

The winter cluster forms at the first heavy frost, and the winter rest begins. A small stock will winter well on six combs about 14 inches wide by 12 inches deep. Stores should occupy about half the comb area — if there is too much comb the stores are less compact and the winter cluster must move about more. If a small cluster loses contact for long with the outside combs it may well starve. (A small cluster will span only three combs in the lowest temperatures).

After mid-winter, when the days start to lengthen, brood rearing restarts diffidently, using stored honey and pollen. The small cluster of the young colony will provide inadequate space within the shell of bees. As the cluster expands, the thickness of its "shell" reduces and heat from the brood area is lost. A small colony will consume more stores than one which has more bees for maintaining adequate insulation around the breeding volume.

State 4. Nest Expansion.

As soon as the bees can fly, fresh pollen is brought in, and brood rearing accelerates.

Sustained by its growing numbers of young bees, the new colony will strive towards full development. It has come out of winter with less than half the comb area of a mature colony. Surplus young bees, for whom there is no space on the combs, cluster near the entrance and build perhaps three new combs by mid to end May.

Comb building absorbs both surplus stores and bee strength. The colony must still be single minded to ensure survival, and there is no urge yet to reproduce.

State 5. Provisioning.

With its nine combs the colony can build a powerful foraging force on the spring flow.

To store the summer harvest, the colony may build two more storage (drone) combs, and it will deepen and widen its earlier nest with drone cells. (By now, the central combs are strengthened by several layers of pupa cases in all the cells right from the top of the comb).

The colony is attaining full development, and it is stocking up not only for the winter but also in preparation for the effort of reproducing itself the following spring.

By the end of the Summer, the colony may have a full set (eleven) of combs, now perhaps 18" deep by 14" wide. The total area of cells will therefore be some 2,500 sq. inches, of which half will be sufficient to store 60lbs of surplus stores over winter.

The stores will be packed in an arch over the late summer brood nest, and will extend deep down in the outer combs on each side.

Supercedure

If the swarm queen was not superceeded by the end of the first season, it is likely to be superceeded now. The colony is maturing in order to swarm the next season, and it must be rare for any queen to be strong enough to swarm once, found a new colony, and swarm a second time.

Generally, each successful swarming will be followed by supercedure.

State 6. Mature Wintering.

About eleven combs is probably the maximum for good wintering. Even a very large cluster will not cover more than six combs in cold weather, and unless the stores are accessible whenever the cluster can expand there will be danger in the critical spring months.

After winter's turn, the mature colony can rear brood, almost irrespective of extreme cold. It has sufficient bees to insulate the breeding volume, and there are stores in plenty including pollen. The flanking combs of stores also play an important role in providing heat capacity at the sides of the brood nest, so shielding the cluster from sudden drops in external temperature.

State 7. Colony Expansion.

Young bees will hatch from January onwards. The colony now has its "skeleton" complete (the combs) and the burgeoning strength is used only to build up the population. By early May, the brood nest may stretch across seven combs, and during May it will double in volume.

Brood nest expansion will run through four cycles between late January and the end of May — the brood taking three weeks to mature in each cycle before the cells can be re-used. The nest swells each time. The final expansion brings the drone cells at the edges of the combs into the breeding area, and a strong colony hatches hundreds of drones by late May.

State 8. Capability of Swarming.

With the hatching of drones in quantity, the colony enters the state when it can reproduce itself. To achieve this, young queens are reared to be mated by drones.

Before the virgins hatch, it is necessary for the queen to leave the nest in a swarm, for it is natural for honey bee queens (being so specialised as to be dependant on the workers) to fight as rivals for the resources of the nest. The bees restrict the virgins to their cradles until the mother has vacated the nest, and release one virgin at a time.

Throughout the early expansion of the brood nest the emerging bees were absorbed into nest duties. As the rate of expansion slows in May, more bees hatch than are needed to service the fully developed brood nest. Unoccupied young bees cluster near the entrance. This is the surplus strength that is used to form the swarm. Whilst not every colony in this state will swarm, it has been estimated by Seeley and Morse that mature colonies have a natural urge to swarm each year unless weakened by disease or mismanagement.

The behavioural sequence which preceeds the emergence of a swarm is complex. A number of events are interlinked:

> the raising of drones
> the building of queen cups
> egg laying in queen cups
> the raising of queen larvae
> the sealing of queen cells
> the decision to swarm, or to break down
>> the queen cells
>
> the giving of the signal to swarm out
> the emergence of the first virgin
> the decision on whether to cast, or to break
>> down remaining cells

For beekeeping purposes we should note:

a) Eggs are rarely, (perhaps never) laid in queen cups before the first drones are hatched.

b) The colony decides whether to throw a swarm or not in the period between the sealing of the first queen cell and its hatching. Once a colony has sealed cells it can only be reliably dissuaded from swarming by being artificially swarmed within the hive.

c) The decision on whether to throw a cast or not has been taken before the emergence of each successive virgin. The bees consider the residual strength of the colony, and if it has been artificially weakened by removal of flying bees only one cell will be allowed to mature.

State 9. Re-stocking.

If the colony swarms, the parent stock is later headed by a virgin queen, which must mate before becoming the new queen of the colony.

Swarming therefore causes a break in egg laying of at least 2½ weeks. In consequence, whilst the population increases during the first 3 weeks as the old brood hatches, the colony thereafter dwindles.

the urge after swarming is to husband strength and to restock for swarming the next season, − so a colony will not swarm later in the same season with the new queen however strong the colony may become.

204. **The biennial cycle in an established colony.**

A mature colony continues year by year in the same cycle: **restocking/ colony expansion/swarming**. In some years, the colony will not swarm if stores are low and numbers of surplus bees have not built up.

205. At some time the cycle breaks when the colony dies from disease or natural disaster. In all probability a fresh swarm will reoccupy the nest site, but this may be only after moth and mice have broken down the old combs, (so clearing a possible reservoir of disease).

Chapter 3

REFLECTIONS

301. Beekeeping is usually a matter of managing full-sized colonies which have been artificially assisted (with feed and foundation) to mature quickly. A mature colony, cannot easily be restrained from swarming; and there is a plethora of "trial and error" systems in the literature, none of which is recognised as universally successful.

Even after a century of trial and experiment, there is still need to find a system which is reliable, convenient and economical for keeping bees in suburban gardens.

302. It is now generally accepted that the whole colony forms one biological organism. The queen, the field bees and nurse bees, the virgin queens and the drones are as dependent on each other as the leaves, the root, and the flowers of a single plant.

Any theory on beekeeping must consider the whole colony as an entity.

303. Honey bees evolved many millions of years ago (whereas in contrast, man emerged only two hundred thousand years ago). The biological urge of honey bees is to produce offspring (swarms and casts). The beekeeper's ideal of a strain which supersedes its queens consistently without ever swarming would, in evolutionary terms, be both sterile and decadent.

We should keep strong, virile strains which have a strong will to reproduce naturally.

304. New honey bee colonies appear to develop over two seasons before the first attempts to swarm.

The first 12 months are spent in founding and enlarging the nest. The next 12 months (from summer to summer) are spent on provisioning the nest and building up the population for the effort of reproduction.

The urge for storage of honey seems to be not so much for survival over winter, but for reproduction the next year. Colonies will store 200lbs or more if nectar is abundant, whereas the bees need only 40 to 90lbs for winter.

305. The maturing of a new colony over two years is clearly a "programmed" sequence. Cohesion of purpose at each step is achieved amongst all members of the colony (varying from 20,000 to 60,000).

It is likely a great number of chemical "messages" (pheremones) circulate which emanate from the queen, brood, drones, – and are transferred from bee to bee or imprinted into the comb. The few distinct pheremones which had so far been identified by scientists were described by Free (1977), but the presence of many more must be suspected.

306. The final transition from **colony expansion** into **swarming** does not occur as inevitably or rigidly in honey bees as it does in, say, bumble bees. Bumble bees always rear young queens in late summer in every year — they must, for the queen lives for only one active season. Honey bee queens in contrast have developed longevity, in conjunction with the evolution of durable nests of wax. As a result honey bees are not forced to swarm every year if conditions are not suitable.

   A mature colony will therefore not always prepare for **swarming** after **colony expansion**, and it may fluctuate between the states of **swarming** and **provisioning**. However, every strong colony must be expected to attempt **swarming**.

307. The complex chain of messages relating to reproduction is not yet fully known. It may start when drones first emerge, (or even while the first drone larvae are being fed). If it is the rearing of drones that ushers in the fertile period, and we could block the impulses to rear drones, a colony might never swarm.

   Free (1977) states that colonies with less than 6000 bees do not build drone cells. If that is due to a directly acting pheremone, there may be the perfect "birth control drug" for honey bees! No doubt experimentation is continuing, but it may be that small colonies lack some natural impetus to rear drones, which cannot be suppressed in a mature colony.

308. It has been proposed by A.E. McArthur (1984) that swarming can be stopped at any time before queen cells are sealed solely by requeening with a new (current year) queen. He depends on mating new queens from nuclei in early May, — to be ready for the earliest swarm preparations. In many seasons, queens fail to mate so early. For this reason over the last fifty years, Bro. Adam (1975) has maintained special nuclei year round, introducing overwintered but unworked queens to the production colonies in March. It has been found by Bro. Adam however that some colonies (though not many) still prepare to swarm, and so the swarming propensity may be affected by the age of the queen and not only by the amount of work it has done.

309. The theory of 'states-of-mind' can be used to support a method of management which also relies on requeening with a current year queen. The start of swarming should first be delayed to mid June through re-cycling the colony expansion. This is easily effected with the deep brood frames in the Long Deep Hive. This hive also facilitates the making of an artificial swarm (to run the colony through to the end of its programmed sequence), should any colony be found with advanced queen cells.

310. The manipulations are set out in Chapter 5. The basic principles are simple (**Method 1**):

    A. Each colony is reduced to about nine deep combs after the main flow — and fed in September to fill half the comb area.

        The colony, although it still has a large population, is thereby regressed from **mature wintering** to **immature wintering**, when it cannot resist and must adjust. (Feed too early and it will re-build the stolen comb!)

    B. Next spring, the colony enters **nest expansion**, accepting frames of foundation. The colony will race along (with its over-wintered population) and extra storage space should be provided in built comb. The building of new brood combs should nevertheless delay the onset of swarming.

    C. A strong colony is left no longer than late May. It must then be kept within the state of **colony expansion** by transferring the brood frames to the rear of the hive and inducing the colony to develop a second brood nest on drawn combs at the front of the hive. The colony may be left for three weeks (to mid-June) to fill six new combs with brood before there will be any danger again of swarming.

    D. The nurse bees on the brood frames which were transferred to the rear will usually raise queen cells (under the supersedure impulse) being at such a distance from the queen that their supply of queen-presence pheromone is reduced.

        A week after transfer, a nucleus with a sealed queen cell can be formed against the rear entrance of the hive. The nucleus is made totally separate with a dividing dummy. It will raise a new laying queen by late June (four weeks from the original brood transfer).

    E. At mid June, the old queen may be caged for one week and then removed. The colony is immediately requeened from the nucleus, using a slow introduction cage. Once it is headed by a young newly-mated queen, the colony will have been manipulated into the state of a stock which has completed **swarming** for that season, and it will start **re-stocking** for the following year.

Dartington Long-Deep hives for 14x12 frames - and Long-Standard hive for standard British brood frames - at HoneyWorks training apiary, Hitchin.

The 'garden-model' hives have splayed legs and gabled roofs, the 'country-model has straight legs and flat roofs.

Colony and nucleus separated by divider, rear entrance opened

Explanatory note below: This colony was first divided into 'swarm' with old queen on left, and 'parent' with brood on right. When 'parent' had raised queen cells, surplus frames were moved back from 'parent' to strengthen 'swarm', leaving just a 4-frame mating nuc at the rear. When the new queen in the nucleus has been proved, the old queen in the 'swarm' will be removed and the nucleus with new laying queen will be combined simply by removing the divider. The frames will be rearranged at the end of the honey flow to leave the brood on the front 9 frames so that the honey in the rear frames can be extracted.

Long Deep Top-Bar hive dating from 1980, with side observation panel to test response of a swarm to making a nest in a long deep box.

15 top bars occupied by mature colony. 12 inch / 300mm natural brood comb on top-bar. Narrow top bars allow access to a super above.

Since the hive had a laying queen throughout, the population will become enormous and a crop can be obtained in supers (and partly in the hatched out brood combs).

With a Deep Long Hive, the second brood nest and the queen raising nucleus are contained within the one hive body. The only extra equipment needed is a queen excluder frame and the dividing dummy!

311. Will Method 1 always be reliable?

The queen cells in the transferred brood frames are raised under the supersedure impulse, and will be sealed within one week of the transfer (when the nucleus should be made and all but one cell destroyed).

The sealing of supersedure cells will not normally precipitate swarming. If however, the old queen in the second brood nest is failing, and does not start a vigorous second nest, the bees might swarm on the supersedure cells (but I have never experienced this myself without true swarm cells being raised in addition near to the queen).

312. Artificial swarming is very simple in a Long Deep Hive. **(Method II)**. It should be carried out whenever sealed swarm cells are found in the hive. Remove all brood combs (but not the queen) from the front half of the hive and insert the divider on the half way line. The nucleus at the rear will therefore also become one half hive size. Any combs with brood that cannot go into the rear half must be exchanged for an empty comb from another hive.

A refinement is to cage the queen for 24 hrs after rearranging the combs, and only insert the divider when the bees have separated themselves into "swarm bees" clustered near to the queen, and "stay behind bees" still tending the brood. To slide in the divider and release the queen can be done quietly with very little disturbance.

313. If it is desired to increase the number of stocks, or sell a nucleus, it is of course easy merely to exchange the queens of the main stock and the nucleus in Method I **(Method III)**.

The nucleus will not swarm with the old queen — it will be concerned with extending its few combs and provisioning for winter, but it may well superceed the old queen in the autumn.

314. It should be possible to get larger crops from the heather by keeping the nucleus and the old queen within the parent hive up to the time the hive is transported **(Method IV)**, so increasing the amount of brood raised in summer.

The nucleus should be increased to five frames and given a super. Combs must be exchanged from time to time to save the nucleus becoming overcrowded.

The old queen is removed, and the two brood nests united when the stocks are ready to be moved.

315. Beekeepers in areas with a strong spring flow could experiment with the "stock and a half system". **(Method V)**, which was tried by K.K. Clark (1951). In this system the nucleus is kept totally separate right through the winter and united only in the spring.

A two queen system published by G.Wells was all the vogue at the turn of the century. Wells wintered each pair of stocks on either side of a thin wood divider so that the two clusters formed one large sphere (so saving on heat loss). I would prefer to winter the clusters well separated for fear that the smaller cluster will have insufficient mobility to reach its stores if it clings permanently against the divider.

316. I freely admit that not all stocks develop well enough in spring for Methods I to V to be appropriate. I give two half supers and then leave smaller colonies to say mid June, when the queen is confined behind the divider in a rear nucleus. The main stock will raise emergency queen cells, and one of these is substituted in the nucleus for the old queen, — who is slipped back into the stock until a new queen is ready to be introduced from the nucleus. **Method VI**.

Chapter 4

EQUIPMENT

401. The Long Deep Hive will need 21 deep frames (and two dummies) in the body, and twenty shallow frames in four half supers; it can then store over 120lbs of surplus. A National will need 44 frames in bodies and supers, and can store 110lbs.

The major disadvantage of the National, as explained in paragraph 103, is the lifting of the top brood box weighing 50-60lbs on and off the top of the pile. The operation of a Long Deep Hive requires the lifting of no more than a half-super, weighing 15-20lbs, and only at a convenient height.

402. The parts of the Dartington Long Deep Hive will be described fully in a companion booklet to aid home manufacture.
The essential parts of the hive are:

402A. The Hive Body.

There is a bee space over the frames, and 22mm under the frames for ventilation.

There is a tunnel entrance at each end which can take an entrance block, and each tunnel leads to a slot in the hive floor. The alighting boards are under the hive, between each pair of legs.

The overall size of the hive body would be covered by two standard National supers, if those were used.

402B. Two Entrance Blocks.

The entrance block slides into the tunnel entrance against fixed pieces such that the floor slot cannot be obscured. The space in front of the block provides clear alighting space — and can be easily closed with expanded foam when it is desired to shut in the bees.

402C. Twenty-one BS 14" by 12" Extra Deep Self Spacing Hoffman Frames.

All frames are fitted with worker foundation and will be used in alternate years for brood and storage.

402D. Two Dummy Frames.

These are the same size as standard frames but have hardboard sides and are filled with polystyrene insulation. The bottom bar is one solid piece.

402E. The Excluder Frame.

The frame has lugs so that it hangs leaving only 2mm clearance both under the covers and over the floor.

The Excluder Frame consists of a zinc excluder sheet with the edges bound with metal flats.

402F. The Divider.

The Divider seals against the hive sides, the cover boards and the floor.

When in use, the space under the lugs must be made bee-proof with small blocks.

The Divider is made of plywood 10mm thick.

The edges are rubbed with vaseline before use.

402G. Four Cover Boards

The covers are simple pieces of plywood 20mm thick. The end covers each have an opening to take a standard Porter Bee Escape; the two central covers are plain.

Each cover board covers one half super, and would cover half a National super, if that were used.

402H. Roof.

Any roof can be used which throws rain clear and provides ventilation over the cover boards.

A suitable roof resembles a National roof but is twice as long.

402J. Four Half-Supers.

Each Half-Super will take five super frames, and is half the size of a standard National super. The bee space is at the top.

If preferred, full supers can be used in place of pairs of half-supers. Full supers should be Modified National in construction, but should have top bee space. (I used my old standard bottom bee space supers for years, with only some slight inconvenience.) The great advantage of half-supers is of course in weight, but the greater number needed would not suit the commercially inclined beekeeper.

402K. Twenty BS Shallow Frames.

All super frames should be fitted with worker foundation.

402L. Feeder.

A 3½ gallon (15 litres) fermenting bucket is ideal — and can be used for other purposes 50 weeks in the year! A circular float must be provided of plywood or similar. This should be cut to fit the bottom of the feeder internally. If the float leaves more than 10mm of syrup exposed when the feeder is full, a "skirt" of plastic greenhouse mesh should be fixed under the float.

403. Although not part of the hive, a pair of Carrying Boxes for deep frames will be found very useful with each Long Hive.

Each box should take six frames so that a pair will store the 12 frames added and subtracted to a Long Hive each season.

A Carrying Box is constructed similarly to a deep Half Super, but the walls should be only of 3mm plywood. It should have a bee space both top and bottom. The box, with a flat lid, can be used both as a temporary super (when clearing frames over a bee escape) and also as a nucleus hive for mating spare queens, with a Half-Clearer Board as the floor.

404. The Half-Clearer Boards are made of 3mm plywood with battening around one side 10mm thick.

A pair will be useful when it comes to clearing the Half-Supers. Throughout the season, each board can be clipped to form a floor to each Carrying Box, so that it can act as a nucleus hive if desired.

Chapter 5

OPERATIONS

501. The detailed operation of a fully developed colony in a Long Deep Hive is most easily explained if we start while the cluster "rests" over winter.

502. The following Twelve-Trip Plan will manage a colony through **Method 1**. (310)

Dates can only be approximate so May I means "the first week in May". All months are assumed to have four weeks only. Precise timing must always depend on season and locality.

The codes (for use in the apiary notebook) list the contents of the hive body after each visit:

*= open entrance
d= dummy frame
b= brood frame
f= frame of foundation
c= empty drawn comb
s= storage comb
X= vertical queen eXcluder
D= Divider

The roof is ignored throughout this chapter both in description and diagrams. Front of the hive is to the left of the diagrams.

503. January 1.: Winter

5031. * Check roof;
clear entrances.

* Lift inner covers gently and check the cluster is quiet and undisturbed.

* Cover holes in inner covers.

* Close rear entrance.

5032. Brood rearing will restart while the bees are still in the state of **mature wintering**, and the colony will need to avoid draughts.

If the colony was well fed and has not been disturbed, nothing more needs to be done.

504. April I; Expansion

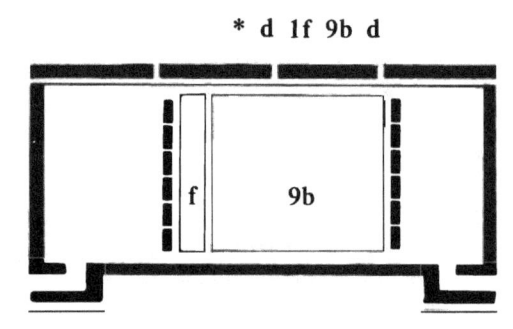

```
        * d 1f 9b d
```

[diagram: hive layout with f and 9b]

5041. * Draw forward the front dummy and add one frame of foundation.

5042. The foundation is best placed in front of all the combs — the bees will then be induced to clear the front frames of old pollen and not ignore those reserves.

5043. Each new comb is drawn out between a flat dummy and the flat front of the former front comb — and will be flat in its turn.

5044. If the colony has empty combs not covered by bees, these should be temporarily stored at the extreme rear of the hive, and put back on later visits, as the colony picks up.

5045. Whenever deep frames are moved, each frame should be "cracked" sideways before lifting (using the square lug on the hive tool). This breaks the propolis without danger of breaking the frame lugs instead. A Long Hive gives plenty of room for manoeuvering frames — which should be twisted within the hive body before lifting out to avoid crushing any bees on the side bars.

505. April III.

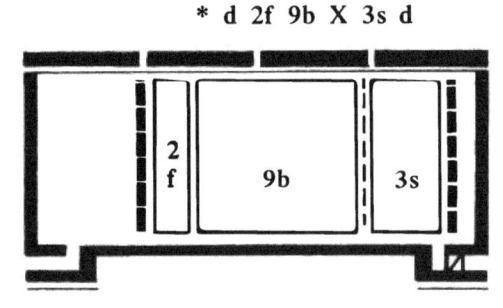

```
        * d 2f 9b X 3s d
```

[diagram: hive layout with 2f, 9b, and 3s]

5051.
- * Draw forward the front dummy and add the second foundation frame.
- * Draw back the rear dummy; add the excluder frame and add three drawn combs for storage;
- * Alternatively, add one half-super.

5052. If the first foundation has not been drawn out, delay giving the second.

5053. The giving of storage space must not be delayed for fear of congesting the brood area.

5054. If the combs were correctly positioned the previous autumn, the vertical queen excluder will lie under the join between the cover boards (The combs must be repositioned if needed).

5055. If the bees can fly to Oilseed Rape, it is probably better to put on two half-supers and no rear storage frames. Bees undoubtedly give first preference to storing over the brood and supers will save Rape honey being stored in the brood frames; the rising heat will also delay crystallisation of that difficult crop.

5056. There does not seem to be a great need for a horizontal queen excluder under supers when 12 inch deep brood frames are used. There is plenty of room in the brood nest and the frame top bars plus airspace is sufficient boundary to the brood area. If the queen does go up it is however essential to fit a horizontal queen excluder once she has been put back below. The most convenient is a WBC unframed zinc excluder, which will lie on the frame bars within the hive walls.

5057. When only one half-super is in use, at least one more should be placed on top of the end covers, so that (although inaccessible to the bees) it will make a level bed for the roof.

506. May I: Supering.

* d 3f 9b X 3s d

5061.
- * Draw forward the front dummy and add the third foundation frame.
- * Add two half-supers (if not already positioned).

5062. The queen now has access to 12 brood frames, which is the maximum she will need for the peak in brood rearing in May. The twelve frame nest has a total capacity of 90,000 cells whereas even a good queen laying at 20,000 eggs per day will use only 42,000 cells for brood.

5063. The supers are essential for a strong stock even if a good spring flow is not expected. Supers help to delay the onset of swarming in two ways: the extra space reduces congestion and the full development of the nest is improved by use of the full depth of the central frames without restriction from any "arch of honey" within the brood frames themselves.

5064. Bees will often run up the side of the first super as soon as it is put on. The second can however be slid down from above so as to scrape back those bees without crushing. A half-super is easy and light to manoeuvre.

5065. There will still be a clear space for three frames in front of the front dummy (and also behind the rear one). Supers cannot be placed over these spaces, or wild comb will be built. If extra supers are needed (for Rape) they must be piled over the central ones.

5066. If the beekeeper is nervous about swarming, a watch can be kept on the front clear space. When a broodnest does not develop steadily because the queen is failing, a cluster of young bees will collect in that space — an incipient swarm cluster.

The cluster will cling to the front cover board when it is lifted; or can be seen easily if a piece of perspex is used to cover the feed hole in the front cover. (The sight is unmistakable, for the bees all hang with their eyes up and motionless).

507. **May III: Transferring brood.**
   * d 3b 6c X 3s 9b d*

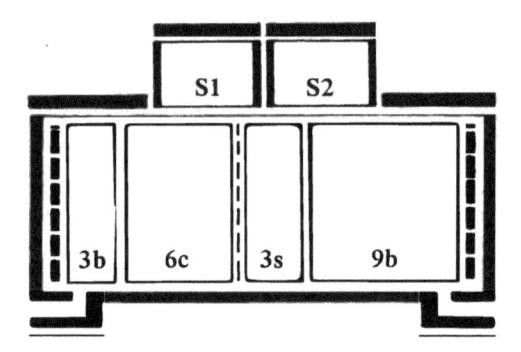

5071. * Lift front cover board; draw the dummy and first three frames to the front of the hive.
   * Lift out the remaining nine brood frames; add six empty drawn combs.
   * Close up the vertical queen excluder and the three store frames.
   * Find the queen and place on the three front brood frames.
   * Replace the nine brood frames at the rear of the hive; close up with the rear dummy.
   * Open rear entrance.

5072. The brood nest is now at its peak and unless action is taken, the colony is very likely to swarm. Substituting empty comb for brood frames makes the bees feel that the brood nest is not yet mature — it returns the colony into **nest expansion**.

5073. The queen must be found in order that she can be put on the front three frames. The beekeeper looks at each frame as he moves it forward or lifts it temporarily out of the hive.

5074. If the weather happens to be bad, it may be preferable to separate the frames into groups so the bees will show where the queen is. In this case, the front three frames should be pulled forward without inspection, and the 9 frames are lifted out into two Carrying Boxes — 4 in one, 5 in the other. The hive is then closed and the boxes taken a little way away (under cover if it is raining).

5075. A glance at the two boxes after perhaps twenty minutes will either show that one group of frames is quiet (and has the queen) and the other is restless, or that both are restless (and the queen is probably in the hive). Where to look is now indicated, and the frames in the Carrying Boxes can be inspected in peace away from the hive. It is only necessary to uncover the rear half when replacing the frames, and that briefly, which causes little disturbance.

5076. The excluder frame will now lie under a super and the queen could climb over. In practice she now has so much room that there is no problem, but if preferred a horizontal excluder can be used as well. It is however essential that the three frames on which the queen is left contain some brood and nurse bees — otherwise I have known the queen to desert the front frames and find her way back to the old nest.

508. **May IV: Forming the Nucleus.**

* d 3b 6c X 7b d D 3s 2b *

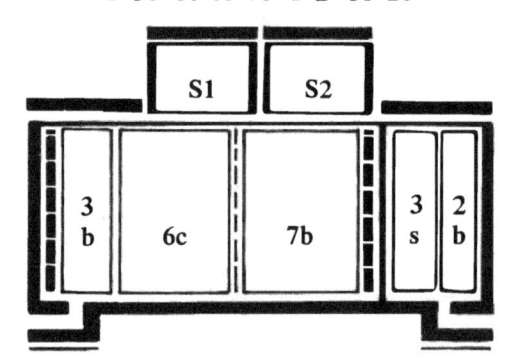

```
         ┌────┐ ┌────┐
         │ S1 │ │ S2 │
┌────┬───┴────┴─┴────┴───┬────┬────┐
│ 3  │                   │ 3  │ 2  │
│ b  │   6c      7b      │ s  │ b  │
│    │                   │    │    │
└────┴───────────────────┴────┴────┘
```

5081. * Inspect each of the 9 transferred brood frames for cells; destroy all cells except one.

   * Form a separate nucleus at the rear of the hive with two brood frames (with cell) and three store combs; close off with the divider; fit a rear entrance block to leave only a small flight hole.

5082. It is best to lift the front cover board briefly; any cluster of bees under the front cover board would mean that the new brood nest is not expanding properly.

5083. Then work under the rear cover boards, one at a time. If the bees are troublesome, lift out the transferred brood frames quietly, and take them away from the hives in Carrying Boxes; the flying bees soon leave and the combs can be checked thoroughly. It is best to work over an old sheet as very young bees will be dislodged which cannot yet fly, and these need shaking back finally. Each part of the frames must be checked, for if even a single queen cell is left it may cause disaster.

5084. If the nucleus is made up with five frames as recommended, it will have sufficient bees to maintain the brood and also sufficient stores. The divider will also seal under the edge of the end cover board. Care must be taken to seal under the top lugs of the divider with small blocks, for if bees can get through then the nucleus will fail to raise a new queen.

5085. The nucleus must not be fed straight away. If the field bees which leave the nucleus to rejoin the main stock take a load of diluted feed with them, they will return and rob out the nucleus.

509. **June II: Checking the new Queen.**

5091. * Inspect the nucleus quietly, opening only the rear cover board.
 * If the weather is poor, add a bottle feeder of diluted honey over the cover board of the nucleus.
 * Find and cage the old queen on the brood frames.

5092. It is three weeks since the brood was transferred; if the new queen was raised from a day old larva, she will now be 10 days old and (if the weather has been fair) about to come in to lay. Check very quietly that the bees seem calm, but otherwise leave alone.

5093. The old queen should now have made a good second nest at the front of the hive. To find her, lift out at least three frames into a Carrying Box and split the remainder into two groups of three each. Inspect each group in turn; if preferred (the number of bees in the hive is now enormous) cover the hive and leave for 10 mins. The group which is then more thickly covered with bees, which are also quieter, has the queen.

5094. Drop a "queen marking cage" over the queen when found and take the frame where you can release the queen on to a closed window. Then cajole her into a "hair curler cage" and cork her in. Replace all frames, and suspend the cage between two frames.

5095. The queen will be fed through the cage walls. No eggs will be laid, but it is now past the "critical date" after which no egg can produce a worker for this year's harvest. The break is beneficial to disease control. The prime purpose is however to ensure that the colony cannot swarm, and also to make sure that the queen can be removed next week however bad the weather or busy the beekeeper.

510. **June III. Re-queening.**

```
        * d 9b X 2b 7c 3s d *
```

S4	S3	S2	S1
9b	2b	7c	3s

5101. * Examine the nucleus for any sign of abnormality.
* Mark and cage the new queen.
* Remove the old queen and introduce the new.
* Remove the separating divider.
* Add two additional half-supers.

5102. The new queen should now be in lay and should be both normal in appearance and have laid a compact patch of eggs.

5103. The new queen can be marked after being confined in the "queen marking cage". Typists white "Typex" is as good a marker as any.

5104. The old queen is removed and the new queen cajoled into the same "hair curler cage". This is now closed with a cork bored with 3/8 (10mm) hole plugged with "queen candy" (one part liquid honey, four parts caster sugar). The cage is replaced in the hive, and the bees will feed the new queen and liberate her within a few hours.

5105. This is classic queen introduction by a very safe method. It may be also that the caging of the old queen (which causes her to stop laying) simulates the old queen going off lay before swarming, and the release of the new queen may simulate the new virgin maturing in a swarmed stock. If so, the method may directly act to trigger the change of state in the colony from **capability of swarming**, to **restocking** — which is our aim.

It might be thought easier to requeen the colony by transferring the new queen complete with the nucleus. In that case however the same combs would remain in the brood nest year after year. In the method described, combs become used for honey storage every second year and so can be sterilised regularly after extraction and before being used again.

5106. The colony is now headed by a queen of the current season raised within the hive — and will not swarm now for the rest of the year. It can be left to gather a crop of honey, partly in the four half-supers (total capacity 60lbs) and partly in the deep frames behind the brood nest (total capacity 96lbs). In practice a crop of 60-90lbs total would be good, and that is achievable only in favourable areas in the UK.

5107. There are now twenty one frames (plus 2 dummies) in the hive body, and twenty frames in the supers. Bees would not occupy all that volume if left to expand horizontally under their own impulses. But bees will not abandon comb that has been brought into use and then transferred as in this case. The super frames should be transferred now — by placing the supers already occupied over the rear half of the hive and the empty ones over the brood.

5108. Colonies managed in this way have been found to be very stable. Such a colony can now be swollen by uniting stray swarms (that have previously been hived in spare supers, checked for disease, and dequeened). The colony can look frighteningly powerful, but does not swarm. I have to remind readers however that the system has not yet been tested comprehensively, with different strains, different locations and in all types of seasons.

5109. When requeening, the beekeeper has more options than Method I. (See Methods III to V, paras **414-416**)' In practice, the Long Hive can be used conveniently for all the numerous methods published for vertical hives. The beekeeper should be able to see for himself a whole range of possibilities.

51010. Method I is the basic system, and demonstrates good practice. Notice that, after requeening, the brood nest is on the nine frames which were provided as foundation or drawn comb earlier that season. If the combs are sterilised each year before use (with acetic acid vapour) the colony should be remarkably healthy, and have the vigour often found in beginner's colonies (where all equipment is new). The hive needs twenty one frames in mid summer, and draws out three foundations each year. The methodical beekeeper can therefore ensure that no combs are more than seven years old — which is again good for disease control.

511. **August III: Removing the Crop.**

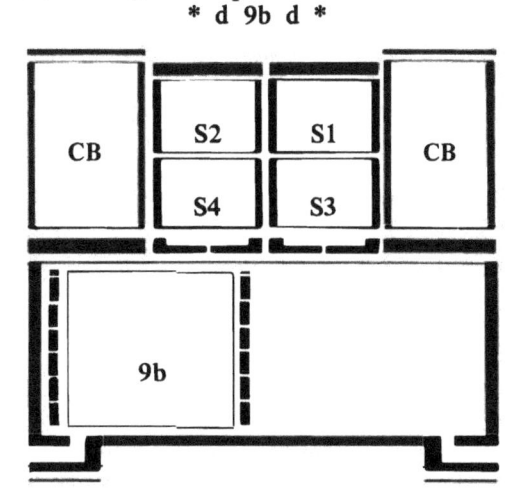

* d 9b d *

CB	S2	S1	CB
	S4	S3	

9b

5111. * Lift off supers.
- * Lift out twelve deep storage frames into two Carrying Boxes; close up rear dummy.
- * Replace the supers over Clearer Boards, placed centrally.
- * Place Carrying Boxes over the end Cover Boards (with a bee escape in each cover).
- * Remove supers and Carrying Boxes when clear of bees (48hours); remove the Clearer Board at the next visit.
- * Fit entrance blocks.

5112. By mid-August in my locality the bees are quiet again after the end of the main flow. A week earlier and bees can still be bad tempered and inclined to rob from frustration as the flow ends. I never take off honey until bees are quiet — which also gives time for the honey to ripen. August Bank Holiday is usually an ideal time for extracting.

5113. If a beekeeper does not have proper Carrying Boxes, it is quite practical (but messy) to use cardboard boxes. The frames are stacked on the ground — then when the hive has been closed each can be brushed free of bees and quietly put away.

512. **September II: Autumn Feeding**.

* d 9b d F *

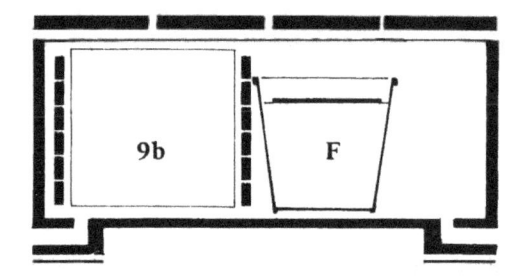

- * Place a full bucket feeder within the hive, touching the rear dummy.

5121. The feeder can be put on the hive floor, but it is better to raise it on two sticks, to avoid crushing bees underneath.

5122. The feeder must touch the rear dummy firmly. The bees then have a short run down into it. The bees first dry up syrup on the walls and then run down the plastic without trouble to drink around the edge of the float. A strong colony will empty three gallons in three days.

513. September IV: Close down.

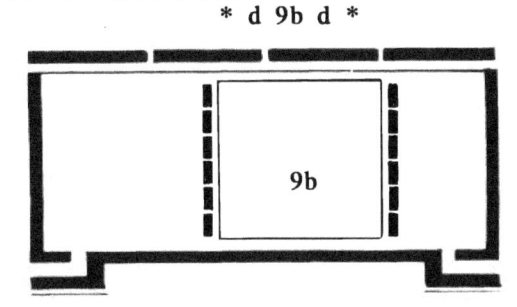

* d 9b d *

5131. * Remove feeder.
* Reposition frames and dummies centrally; the rear dummy should lie just under the second cover board.
* Uncover holes in the end cover boards.

5132. Two weeks after feeding the bees will again have quietened down. The feeder can be removed quietly, together with any wild comb induced by the sudden flood of food.

5133. The combs are positioned centrally to provide plenty of free air for wintering, and so that foundation can be added next spring in the front void without disturbance.

If the rear dummy lies under the end cover board the excluder frame will naturally be correctly positioned when storage frames are again needed.

514. This concludes the twelve visits in the "Twelve-Trip Plan." Beekeeping is rarely so neat — but could be if queens were consistently of good performance and seasons were not too variable.

515. Winter work for the beekeeper should include: sterilising stored combs (in their Carrying Boxes); melting cappings and scrapped combs (using a universally useful "Hot Box" — an insulated box heated by two light bulbs); casting foundation for next year (with a foundation mould); and making equipment.

Chapter 6

CONCLUSIONS

601. It is natural to ask of an unorthodox approach:
- does it work?
- is it really new?
- is it economical?

602. It is ten years since I made my first Deep Long Hives, so they have been used through a variety of seasons together with Nationals, first in an inner London borough, and lately near open country-side. My apiary is now reduced to five long hives and the long observation bar hive.

I do not claim enormous crops, but I get far more honey than I need. The system will need rigorous testing by a 'commercially minded' beekeeper before it can be compared for honey yield with conventional systems.

If the Long Deep Hive is found to yield less, it would imply that the hive proportions cause the bees to use excessive energy over raising the brood or maturing nectar. This could well be so in a long hive holding 30 standard combs; that hive would be six times as long as it is deep. A Long Deep Hive is two and a third times as long as deep until the supers are added, and then the length is only one and three quarters the height. Although I have no scientific evidence, I doubt bees have any great difficulty controlling temperature and ventilation in a Long Deep Hive.

603. There can be no doubt that the operation of a Long Deep Hive is more convenient for the garden beekeeper than National or WBC hives, which must extend the brood nest across two boxes and generally two sizes of frames.

A Long Deep Hive is also more convenient than a simple hive with a large single brood chamber, since separate nucleus hives are then necessary for rearing the replacement queens.

604. It is very easy to use a Long Deep Hive for almost all the systems of management that have been worked out for tiered hives. The normal diagrammatic arrangements of brood, excluders and supers need only to be read 'sideways' instead of 'upwards'. The rear entrance of the Long Deep Hive substitutes for a top entrance whenever that is required.

A Long Deep Hive facilitates all such systems since just as many frames as are needed can be used in each part of the hive, whereas with a tiered hive the manipulations have to be with full boxes of 11 frames.

605. Method I is now the basis of my operations (although I depart from it according to the bees' needs or my shortage of time). It complies with good principles, in that:

* The winter cluster is well ventilated.
* The winter cluster (on 9 deep combs) is compact, and the deep combs do not distort the early brood nest in any way.
* New brood combs are drawn out from foundation in early spring (when perfect combs will be made).
* Empty supers are placed over the active brood nest, where incoming nectar is most readily stored.
* Empty comb is provided for a second brood nest, before the brood nest has been fully developed, so delaying the onset of swarming.
* The second brood nest is developed on combs which have either been drawn out that season or were added from store (after disinfection), so controlling Nosema and other diseases transmitted on comb.
* Swarming is controlled by annual requeening with a current year queen before the colony has started preparations for swarming.
* The replacement queens are reared within the full colony.
* No flying bees are lost during the making and reuniting of mating nuclei.
* Egg laying is maintained at full pitch right up to the 'critical date' (after which an egg will not become a forager in time for that season).
* After the critical date, the number of brood frames available to the queen is restricted to only what is necessary.
* No honey need be removed before it is fully ripe.
* The colony is fed for winter in one go – so that the entire feed is taken just as fast as the bees can manage, and too quickly for an excessive amount to be used up in a late brood nest (which would prevent the proper positioning of winter stores).

606. Not one of the individual techniques used to satisfy each of the principles above is original, and none claims any new discovery about bees.

The theory provides a plausible explanation of why this way of managing bees is successful. Each individual technique has been found by earlier beekeepers mostly by trial and error.

However, so far as I know, no complete system has ever been published which combines all these techniques into one system.

607. To prove the economy of the system will require more exercises than I have yet carried out. The capital costs of equipment need to be compared, together with 'time and motion' studies on beekeepers. The question of comparable honey yields must be settled. Whether any particular beekeeper will find benefits from using Long Deep Hives may well depend on what he thinks most important:

> average yield per stock successfully overwintered;
> highest individual yield in the apiary;
> return per £1 invested;
> return per beekeeper's hour.

In my belief, Long Deep Hives will perform well on 'average yield per stock successfully overwintered' and on 'return per bee keeper's hour'. These seem the important criteria for the garden bee keeper under modern conditions.

But above all, keeping bees in Long Deep Hives is highly enjoyable.

REFERENCES

BROTHER ADAM (1975).	Beekeeping at Buckfast Abbey. British Bee Publications Ltd
LESLIE BAILEY (1981).	Honey Bee Pathology. Academic Press, London
E.R. BENT (1946).	Swarm Control Survey. Gale and Polden Ltd
EDWARD BEVAN (1834).	The Honey Bee, its Natural History, Physiology and Management. London
F.R. CHESHIRE (1886).	Bees and Beekeeping; Scientific and Practical. London
KENNETH K. CLARK (1951).	Beekeeping. Penguin Handbooks
T.W. COWAN (1881)	The Social Organisation of Honey Bees. Edward Arnold, London
W. HERROD-HEMPSALL (1938).	The Beekeeper's Guide. British Bee Journal, London
A.E. McARTHUR (1984)	Milestones in Beekeeping and the Swarm Trigger Discovered. Privately published
DR. C.C. MILLER (1915).	50 Years Among Bees. A.I. Root Company, Ohio
THOMAS D. SEELEY & ROGER A. MORSE (1978).	Bait Hives for Honey Bees. Cornell University, New York
JOHN SHIDA (1976).	Beekeeping. Thornhill Press Ltd, Gloucester

S. SIMMINS (1904).	A Modern Bee Farm. London
L.E. SNELGROVE (1934)	Swarming, its Control and Prevention. Privately published
L.E. SNELGROVE (1940).	The Introduction of Queen Bees. Privately Published
H.J. WADEY (1948).	The Behaviour of Bees and of Beekeepers. Bee Craft
E.B. WEDMORE (1932).	A Manual of Beekeeping for English Speaking Beekeepers. Bee Books New and Old

Electric Extractor

Only 8 ½" High

UK Patent No. 2036526
awaiting publication

Honey-Twin-spin

Developed and manufactured in Great Britain

Brinsea Products